The Snuggle Book

Turnbull & Willoughby
Publishers

The Snuggle Book

Lennie Rose

Illustrated by D.S. Van Cleaf

First printing October, 1985.

10 9 8 7 6 5 4 3 2

Manufactured in the United States of America.

Published by Turnbull & Willoughby Publishers, Inc.
1151 W. Webster, Chicago, Illinois 60614

ISBN: 0-043084-31-8

To: ____________________

From: ____________________

Snuggle Me!!

Introduction

This book is about how to snuggle more in this increasingly snuggle-less world. It will enlarge your snuggling repertoire and enhance your snuggle-ability. You will discover that snuggling reduces headaches, eases tensions, clears complexions and puts a smile on your face.

This is the perfect book for someone you're currently snuggling with, someone you'd like to be snuggling with or someone you know who needs help in snuggling. So relax, get comfortable, turn the pages of The Snuggle Book, reach out and snuggle someone.

The Birth of Snuggle

In prehistoric times, long before the invention of kissing, drive-in movies, and computer dating, a caveman set out on foot in search of a woman. When he found one he liked, he grabbed her by the hair and proceeded to drag her home. During their journey the woman grabbed the club out of her suitor's hand and clobbered him up-side the head. When the caveman awoke, hours later, he found himself comfortably nestled in the arms of his woman. Finding this to be extremely pleasurable, he hit *himself* over the head with his own club for not having thought of it sooner. For a second time he went out cold, only to awaken as before, cuddled in the arms of his woman. Thus the snuggle was born, and shortly thereafter, the aspirin. The headache, however, according to history, was invented months later by the woman.

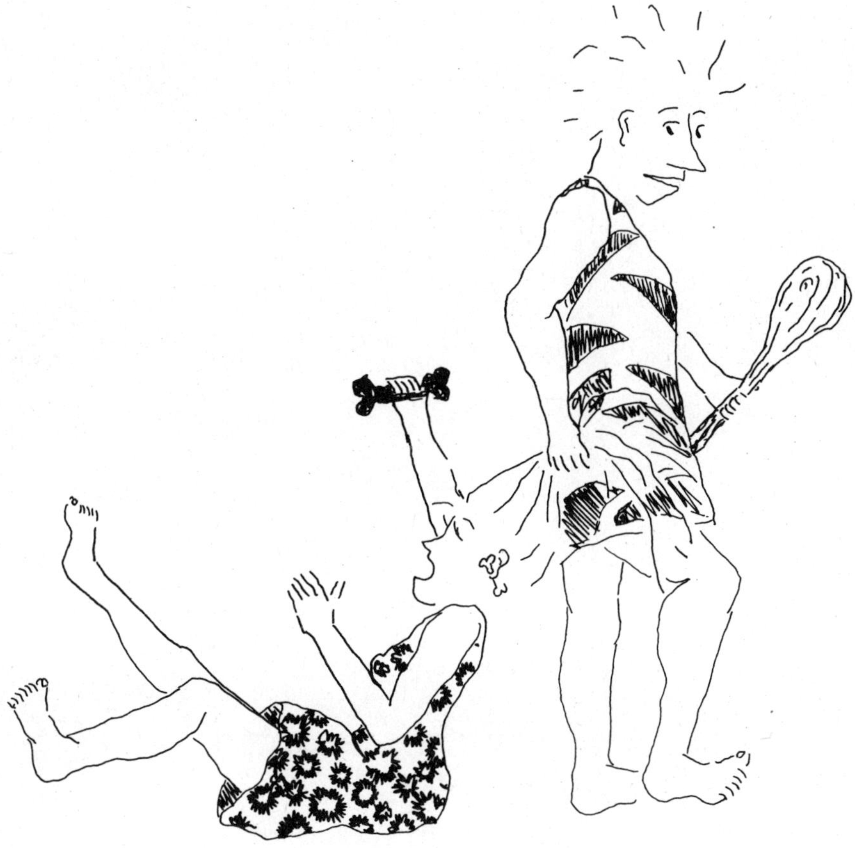

Snuggling is a touching experience.

Section I

PRONE TO SNUGGLE: Horizontal Positions

Snuggling is a 24-hour form of recreation. The snuggles in this section can be done awake or asleep.

The Classic Spoon Snuggle

From time immemorial, snugglers have enjoyed the Classic Spoon Snuggle for it's satisfying comfort and pleasing closeness. Pressed together, front to back, two people are perfectly joined together into one continuous form like two spoons in a drawer. Sweet nothings can be whispered in this position but, for many couples, the cuddling says it all. The Classic Spoon is also wonderful for sleeping.

The Late Night Letterman Snuggle

This snuggle, also called "Flattened Spoons," is a variation on the Classic Spoon position. The rewards are the same but the technique differs. Using less bend at the knees and waist makes this snuggle stretch to meet the needs of watching T.V. on the couch, cuddling on a twin bed or sharing a sleeping bag.

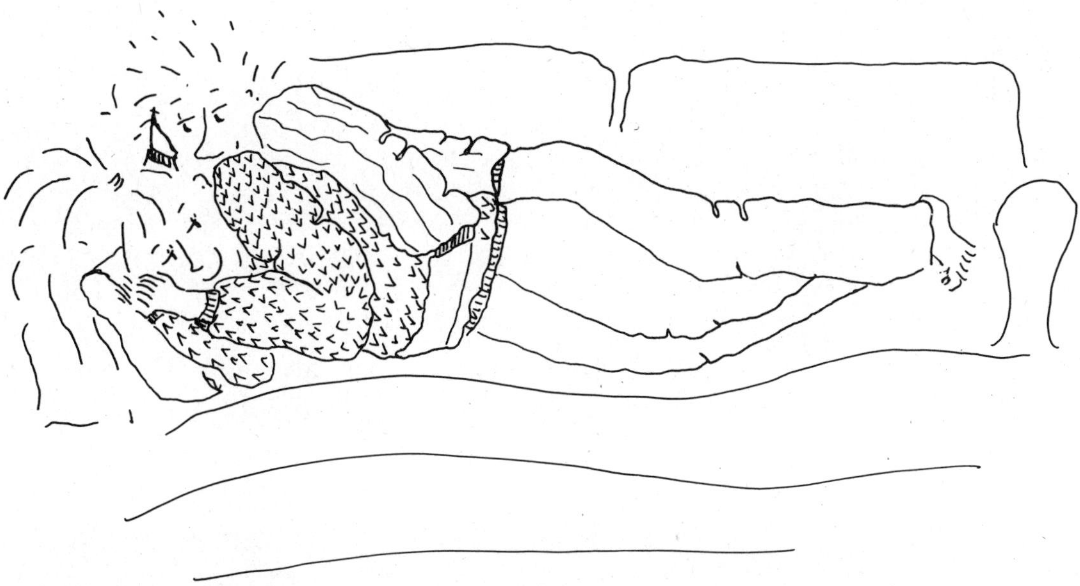

The Hot Crossed Bun Snuggle

The Hot Crossed Bun Snuggle is perfect for pooped people because it takes almost no mental or physical energy. It's the kind of snuggle you should do when you care about someone enough not to ask anything of them but a simple touch. The next time you crawl into bed exhausted, try to simply back up into your Honey's buns, then snuggle there and say goodnight. As you head out into the Land of Nod, you will do so with the quiet reassurance that the person you care for most is nearby.

The Ankle Hook Variation

On a hot August night, when the air conditioner is broken, too much touch can be just too much. But you still want that connection of affection so snuggle up in the Hot Crossed Buns position and hook ankles for that extra cool and comfortable touch.

The Knees and Nose Snuggle

The snuggle can take you by surprise. There you'll be — sound asleep — tossing and turning — and suddenly you'll find yourselves nose to nose and knees to knees. It's hardly ever done on purpose — but if you find yourself waking up in this position, just move over, gently sweep your love up in your arms and snuggle her until you both fall back asleep.

Snuggle in a Moonbeam

Don't wait until the 4th of July to take a blanket out under the stars with your lover. Create your own fireworks with a blanket, a grassy meadow, someone you care for, and the snuggle in a Moonbeam. Lie close and quiet and when you see your first falling star, you may find you have no need to wish for anything more.

Although this snuggle is enhanced by twinkling stars or a full moon, even without romantic scenery it's a perfect "let's have a long talk about how wonderful it is to be together" snuggle.

The Still of the Night Snuggle

This snuggle will make you feel safe, secure, and at peace with the world. Head on shoulder, wrapped together — it can be a comfort in the worst of situations . . . a bad day or a lumpy bed . . . or a celebration of the best of situations: being alone in the still of the night with someone you love.

The Snuggle Explained

A snuggle is entirely different from a simple hug.

A snuggle is a lengthy, mutual, amorous embrace. It's touching with feelings. It offers affection without expectation, real romance and genuine caring. It's cuddling with your hearts as well as your arms; holding and being held. It's serious and playful, romantic and fun.

It's simply the most satisfying way to spend time with someone you care about.

And if someone just gave you this book they deserve one.

MITE

Section II

LOUNGING AROUND: Seated Snuggles

These snuggles can be friendly or romantic depending on your mood, your snugglemate, and your surroundings.

In the silence of each other's arms a thousand words are spoken.

The Bus Stop Snuggle

If the bus is late and you have to wait, seize the opportunity for a snuggle. It's perfectly acceptable in any public situation (stadiums, concerts, park benches or the back seat of your Chevy). This snuggle is a simple way to say how good you feel together. Wherever you find yourselves sitting together, remember to snuggle. It can be habit forming.

If you have just started caring for someone and snuggling diligently is still in the future, this is a perfect way to subtly announce your intentions.

Variation: THE BUS STOP SNUGGLE FOR OLD FRIENDS

When you've had a little more time to get wrapped up in each other, try this variation on The Bus Stop. Why use only head and shoulders when there's someone near you who could really use an all over snuggle?

Variation: THE BUS STOP SNUGGLE FOR OLD, OLD FRIENDS

You're *never* too old to snuggle.

Blow life into that fire and rekindle it with care by snuggling with your sweetheart whether at a bus stop or on the old front porch and together in each other's arms, talk about how you snuggled way back when.

The Lazy Boy Snuggle

There's nothing so comfortable as relaxing in the old easy chair. There's also nothing more surprising than being joined by the roving snuggler who lands in your lap! This unexpected pleasure will add a great deal of zip to what might have been a plain, ordinary moment. It's a romantically playful way to refresh your affections.

The Opposite Ends Snuggle

After a long, busy day, this is one way to make ends meet. Take time to stretch out, talk about what's on your mind and in your heart.

It's a snuggle with less emphasis on physical contact than conversational contact. It's the perfect mid-range snuggle: your eyes are not close enough to require deep expressions of love and yet you're not sitting across the room in silence.

"Remember our first date?" "Maybe we should spend a long weekend together". "You know why I like being with you?"

Serious snugglers will add a foot and leg rub to the Opposite Ends Snuggle.

Variations on Opposite Ends

THE CHIP-AND-DIP SNUGGLE

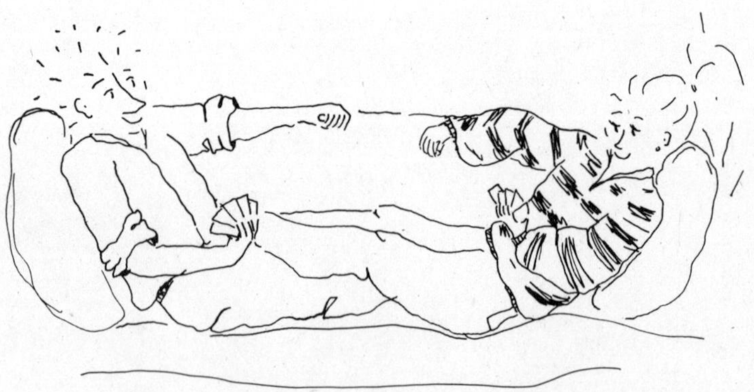

THE GIN RUMMY SNUGGLE

THE SUNDAY PAPER SNUGGLE

THE ALL FOR ONE SNUGGLE

The Daddy Long Legs Snuggle

Does this look uncomfortable? Well it's marvelous!! It's the ultimate "you do something, I'll relax" snuggle.

Dim the lights, bring out the coffee and Cognac for your snugglemate and relax . . . soon they will too because it's easier to reach for you than the coffee.

The Secret Snuggle

Picture yourself in an elegant restaurant. The waiters are efficient. The food is fine. And the one you love to snuggle is across the table. Carefully, remove your shoe and with your stocking foot, reach up and give a little snuggle under the table.

Word of caution: If you're at a table for four, be careful who you snuggle.

Snuggler's Blues: The Hazards of Snuggling

Here are a few things that could put a snag in your snuggling and a cramp in your cuddling:

Hands, feet, legs or arms that fall asleep.

Clothing that is too tight.

Major wrinkles (snuggle in your dress clothes only after you've been where you're going.)

Junk jewelry (dangling earrings, pointed plastic fruit, multiple metal bracelets and jumbo belt buckles.)

Urgent phone calls, unwanted visitors and most of all beware of snuggling with someone you don't care for. You can fake a kiss, you can fake a hug, but you can't fake a snuggle.

Section III

SNUGGLE ASSORTMENT: Mixed Snuggles

As you become more accustomed to snuggling, you'll want to try these versatile positions. They are fun in almost any situation and can be done at home or in the great outdoors.

Snuggling is a way of letting yourself go by holding on to someone else.

The Tab A and Slot B Snuggle

When you bought the Snuggle Book you asked yourself how many ways can two arms and two legs possibly go? Well, here's a snuggle that will amaze you with its comfort. This is the friendliest snuggle in the book because it creates a perfect climate for creative conversation. It's an unusual snuggle that will make the two of you feel like an Origami sculpture folding into each other's corners until the position suddenly slips into place.

Make sure drinks and reading materials are handy because once you get comfortable you'll want to stay in this position all afternoon.

Disassembling the Tab A and Slot B is easy; just count to three and "roll out."

(Caution: Be sure to roll in opposite directions!)

The Tab A and Slot B for Beginners

Not everyone is mechanically minded, and if you are not, or if you're not quite as flexible as you used to be, try this simplified version of Tab A and Slot B.

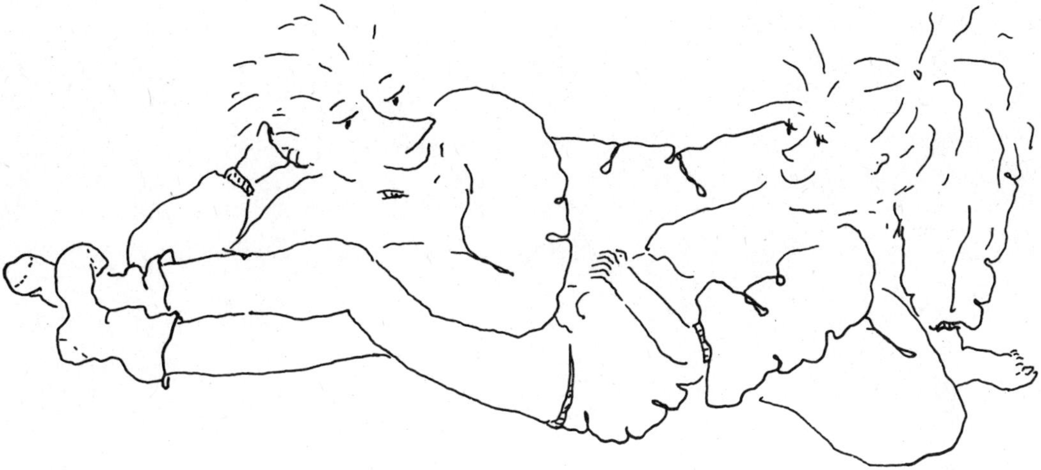

Laplander Part I

This snuggle is famous for its spontaneity. Your partner is already relaxed on the couch, probably watching T.V. Suddenly, you get the urge to snuggle. So slowly, and without a sound, you glide in his direction, slip in beside him, and ease your head on his lap.

Laplander Part II

When you have your snugglemate's full attention, you're ready for Part II. Through the ages, great romantic couples have chosen this position — not only for looking deep into each other's eyes and for feeding grapes but for getting a whole new perspective on their relationship.

Note of Caution: If the person whose head is on your lap goes to sleep, no doubt your lap will too.

Trivial Pursuits Snuggle

Snuggling is never a waste of time. But with the Trivial Pursuit Snuggle, you can accomplish two things at once. Novels, homework, business reports, letters from your aunt — all can be appreciated in this position.

(Note; even though this snuggle is called "trivial pursuits" — the pursuit of snuggling is never trivial).

Why You Should Snuggle

Snuggling is one of the few things in life that is good for you without being bad for you. It's like chocolate without calories, shopping without bills, and champagne without hangovers. Snuggling feels good. It enhances relationships, deepens communication, and puts a smile on your face. It's soothing, nurturing, calming and comforting.

Snuggling adds romance to your life. You can't get a bad reputation from heavy snuggling and the only socially communicable disease you can catch is affection. Snuggling is also a perfect antidote for loneliness because it's virtually impossible to do alone.

Who Should Snuggle

Firemen should snuggle. So should policemen, nurses and cowboys. Also, accountants, copywriters, mechanics, businessmen and women, elevator operators, dentists, doctors, football players, weight lifters, bartenders, beauticians, artists, high school teachers, carpenters, nuclear physicists, and bio chemists.

And don't forget farmers, stockbrokers, housewives, butchers, bakers and candlestick makers and, most of all, ______________________ .

Section IV

STANDING ROOM ONLY: Upright Snuggles

Don't just stand there —
snuggle someone!

The best way to warm yourself up is by lighting a fire in the heart of someone else.

Face to Face Snuggle

At first glance, this may seem more like a hug than a snuggle. But the difference is, if you're wit[h] someone you think is wonderful, you can enjoy this position for a long, long time.

A Face to Face snuggle makes your hellos last longer and your goodbyes more bearable.

Use it whenever you need to make boring moments better and good moments grand.

SHOWER ME WITH AFFECTION SNUGGLE

REACH OUT & TOUCH SOMEONE SNUGGLE

THE LONG GOODBYE

The Snuggle Habit

Cultivate the habit of snuggling anytime of the day or night. Get in the habit of reaching out whenever that person you care about comes near.

Almost everyone gets into the snuggling mood when they're near a cozy fireplace or under a breathtaking full moon, but the true snuggler will find even the flimsiest excuse to hold their loved one near.

The Wraparound Snuggle

This is a snuggle that can enhance sharing a special moment in your lives or make an otherwise mundane moment extra special. Do it when you're stuck standing somewhere pressed together in a crowd, in an elevator, in line at the movies or in a jam-packed restaurant waiting room or after dinner as you wait to get your car from valet parking. You can share a snuggle and go home warmer and happier.

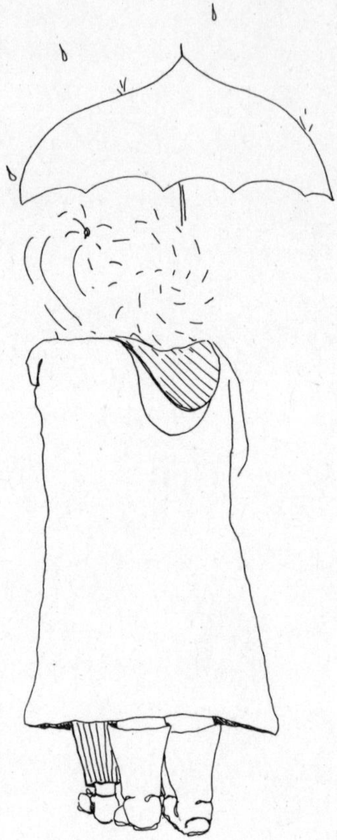

There's nothing as romantic as a wraparound snuggle in the rain under a big old umbrella. Try it at night, cuddling close while you watch the drops roll off the tip of your umbrella and the sky as it lights up around you.

. . . or during a walk through the zoo or proudly watching the kids at a Little League game or anywhere two lovers stroll.

The Off the Wall Snuggle

The Off the Wall Snuggle is the most spontaneous of all the snuggle positions. It expresses a certain heated excitement and passionate sharing of news or affection that can't wait. Often it's fueled by desire and is always done on the spur of the moment. It may take the form of a special welcome home because you couldn't wait to see him, or because he sees you leaning against a wall and can't resist you. You will probably talk about how crazy you are for each other, or how crazy you are for standing there snuggling.

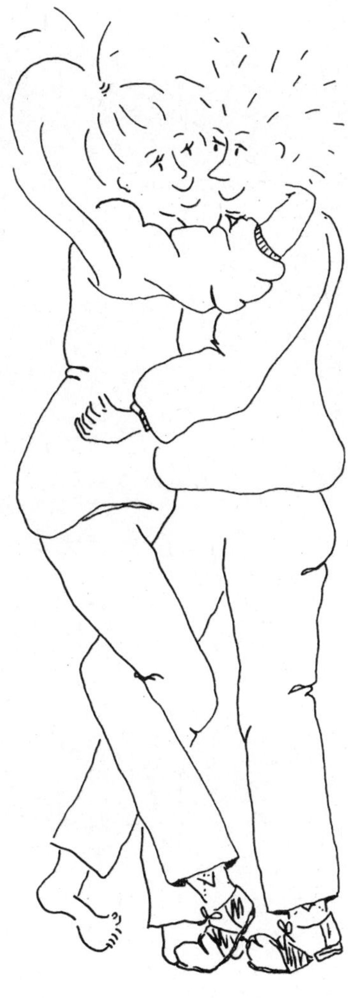

Two for the Road Snuggle

This is a snuggle that allows you to take your feelings on the road. It's a portable snuggle that requires no props — no planning ahead — and it lets people know — wherever you are — that you belong together. Don't leave home without it.

CH
RS

With a little practice, this position can evolve into something permanent.

Available at fine book stores everywhere or order direct from the publisher by using the coupon below.

Turnbull & Willoughby Publishers
1151 W. Webster
Chicago, Illinois 60614

The Snuggle Book

To order one copy simply enclose $5.95 plus $1.25 shipping and handling. For order of 5 or more copies, publisher pays all mailing and handling charges.

Illinois residents add 5% state sales tax.

Enclosed is $ ______________ plus $ ______________ shipping and handling.

Please send ______________ copies.

Mailing Label — Please print clearly

____
Name
____
Address
____
City State Zip